Solving Diophantine Problems

All rights reserved © 2023 by John A. Cramer

No parts of this book may be reproduced or transmitted in any form or by any means, graphic, electronic, or mechanical, including photocopying, recording, taping, or by any information storage retrieval system, without permission in writing from the publisher.

ISBN 978-1-329-96380-1

Printed in the United States

Solving Diophantine Problems

John A. Cramer
Emeritus Professor of Physics
Oglethorpe University, Atlanta, GA

Table of Content

Introduction 1

Chapter 1. Linear Equations..7

   Equations in Two Unknowns...................................7

   Equations in More than Three Unknowns...................14

   Simple Quadratics..20

   Particular Cases of Equation 4...............................25

   Solving Pell's equation...28

   Pythagorean triples..33

   Pythagorean Kin...36

Chapter 3. Useful Lemmas for Higher Orders.................41

   Lemmas on Primes and Square Numbers.................41

   Lemmas on Irrational Numbers...............................44

Chapter 4. Cubic Equations in Two or More Unknowns.....49

   Simple Cubics..49

Chapter 5. Other Equations in Two or More Unknowns........55

   Equations with Inverse Powers................................55

   Systems of Linear Equations...................................60

   Systems of Double Equations..................................65

   Various Sets of Simultaneous Equatio.....................69

   Simultaneous Equation of Higher Order...................72

Solving Diophantine Problems

Introduction

Diophantos of Alexandria was a mid-third century (A.D.) Greek mathematician. Only six books of his most important work, *Arithmatika*, have long been known though he claimed to have written thirteen. Either he never finished the work or some books were lost very early. Hypatia of Alexandria, only a century and a half after him, seems to have known only six books, books 1, 2 and 3 and also three books of, at least now, uncertain numbers in Greek. Recently, volumes 4 through 7 were found, misfiled in the Astan Quds Library. They had been translated from Greek into Arabic by the ninth century Christian scholar Kostas Luca (Qusta ibn Luqa). Thus, we now have seven of the putative thirteen books in hand.

Living near the end of the classical Greek era, Diophantos was heir to both its great benefits and its disadvantages. By that time, Greek mathematics included the great works of Euclid and Ptolemy (also of Alexandria) but it was encumbered with a poor selection of available mathematical symbols. It had no plus sign, no equals sign and no decimal number system. Greek letters were used in place of numbers. Equations as we know them could not be written and had to be verbally stated rather than written symbolically. To his credit, Diophantos began the needed process of inventing algebraic symbols, using an abbreviation for "equals" rather than the word itself.

Furthermore, 1 was *the* unit and zero was nothing at all; so neither was regarded as a number. Negative numbers and imaginary numbers were nonsense, rejected as solutions of the equations. Though they had been recognized centuries earlier, Diophantos avoided irrational numbers.

Today's Diophantine problems are algebraic problems for which integer solutions are required for one or, more challengingly, two or three and possibly even four or more unknowns. Additionally, they are typically word problems, that great bane of students. Following the modern tendencies, I will treat them primarily as bare equations but, in order that we retain a sense of them as they originally were, I will on occasion give an example as a fully stated word problem.

At the time, even equations in a single unknown were a challenge and many such problems appear in *Arithmatika*. Since the typical Diophantine problem sets up just a single equation in the unknowns, Diophantine problems are usually under-determined although the requirement of allowing integer solutions only is a most important additional constraint on the solutions. Some authors (including Diophantos himself) relax the constraint to merely rational solutions but the distinction lacks any great significance.

Thus, although Diophantine problems are fundamentally just algebra problems, their under-determined character, derived from having too many unknowns, means that finding a solution does not fully answer the problem posed. Is the solution the only one or even the best or most relevant one? How many solutions does the problem actually have? If no solution is found, is that because there is none? Might the failure be due to inadequate mathematics? Obviously, Diophantine problems are not just simple algebra and they have, and historically have demonstrated, a penchant for forcing inquirers to probe into mathematical depths. Accordingly, it is good to have in mind, from the outset, a plan for managing them and in this discourse I will follow something like the following schema.

    I. Find a solution
        a) If a solution exists, try to find more
           i) Show they are infinite in number
           ii) Failing that, show they are finite in number
        b) If no other solutions are found, show the solution is unique
    2. If no solution is found, show none exist.

Finding just any solution is not necessarily difficult because trivial solutions are often obvious. By trivial solutions I mean solutions where only one unknown has a non-zero value. Note that, since zero was not a number to Diophantos, he avoided these trivial solutions. Zero values are often physically 3uninteresting as, for example, a right triangle with a side of zero length, but they may nonetheless have mathematical use or interest. In particular, a solution might be used to generate others. Sadly, however, trivial solutions often do not lend themselves to generating more solutions.

Both the finding of a non-trivial solution and then finding more solutions have historically been largely a matter of transforming the equation and/or the solutions with clever algebraic identities. The most famous of these, and probably the earliest, is *Brahmagupta's identity* dating from about 628 A.D.

$$BI \quad (a^2 + fb^2)(c^2 + fd^2) = (ac \pm fbd)^2 + f(ad \mp bc)^2$$

where f is a constant and a, b, c and d are unknowns. It implies the important conclusion that numbers of the form x + fy are a unit ring with the additive identity (0, 0) and multiplicative identity (1, 0). Since x and y are integers, elements have no multiplicative inverse. That is, these numbers form a multiplicative domain.

A transformation I will use often in this monograph, is to "eliminate" some of the unknowns by setting them equal to the first unknown plus or minus an also unknown integral number. Any equation can thus be reduced to an equation in the first unknown (plus a number of unknown "constants"). Then of course ways must be found of coping with the constants.

A method pioneered by the great amateur Pierre Fermat is a variant of proof by contradiction he called the method of infinite descent. He used it to prove the non-existence of solutions rather than to find them so I will not show it although I certainly will give examples of proof by contradiction.

Unfortunately, there is little method in the *Arithmetika*. The solution of one problem is seldom much help for solving the next. Then too, Diophantos had an irritating tendency to state a problem and then present the solution with little or nothing by way of explanation of how he arrived at it. To a modern reader, the original text is maddeningly prolix. This is largely due to the lack of simplifying symbols and systematic terminology for which Diophantos cannot be blamed. Regardless of its faults, the *Arithmetika* has been a most important, perhaps even *the* most important, work ever written for stimulating further mathematical research.

We have no reliable information about Diophantos personally. All we have is contained in a Diophantine word problem in his honor. It was evidently created by Metrodorus in the 5th century.

> Here lies Diophantos, the wonder behold.
> Through art algebraic, the stone tells how old:
> God gave him his boyhood one-sixth of his life,
> One twelfth more as youth while whiskers grew rife;
> And then yet one-seventh ere marriage begun;
> In five years there came a bouncing new son.
> Alas, the dear child of master and sage
> After attaining half the measure of his father's life
> Chill fate took him. After consoling his fate by the
> Science of numbers for four years, he ended his life.

Expressed as an equation where x is his life span, the problem is then

$$x = x/6 + x/12 + x/7 + 5 + x/2 + 4$$

and the solution is that he lived 84 years. Tidy as all this information is, its historicity cannot be verified.

Chapter 1. Linear Equations

Equations in Two Unknowns

Linear equations in a single unknown are no great challenge to solve. Though they can certainly be Diophantine, I will trust the reader to have no need of help with them. Algebra has, after all, progressed considerably since the time of Diophantos.

The above problem on the age of Diophantos is an example but, before leaving the topic, I will give another such problem typical of Diophantine style. It was given to my uncle, Paul Glick, in 8th grade to keep him busy and out of the teacher's hair. Forever after, he was proud that he solved it quickly in his head. Uncle Paul is long gone but I still remember him and his pride in solving the problem. That is the pleasure of finding Diophantine solutions. I hope it bring you Paul's pleasure. I supply the answer at the end of this chapter.

> Paul's Problem: A goose in the barnyard looked up and saw a V of wild geese flying by. He called up to them, "Come down, ye 100 geese." The flight leader looked down and replied, "We are not 100 geese but if we were twice what we are plus one half what we are plus one fourth what we are plus you, we would be 100." How many geese were they?

Linear equations in two unknowns, say x and y, are the next easiest Diophantine problems to solve. They can always be cast into the form

$$ax + by = c \qquad 1)$$

where a, b and c are integers (including negatives and c = zero). These equations are readily solved by assuming y = x + d which then quickly leads to the results that

$$x = \frac{c - bd}{a + b} \quad \text{and} \quad y = \frac{c + ad}{a + b}$$

Trivial Solutions:
　Linear equations of the form of Eq. 1, where a, b and c are known integers, obviously have trivial, rational solutions (where either x or y is zero) of the form (c/a, 0) and/or (0, c/b).

General Solutions:
　A simpler solution format can be found when we substitute an unknown constant m for d by making d = (c - m(a – b))/b in the above general solutions. We find that these linear equations have solutions of the more compact form

$$x = m \quad \text{and} \quad y = \frac{c - ma}{b}$$

We of course require m to be an integer. Hence, every Diophantine linear equation of the above form (eq. 1) has general, non-trivial, rational solutions. Neither these general solutions nor the trivial solutions are certain to be integer solutions but it is not difficult to generate integer solutions from them. It is usually a simple matter to find an integer value of m for which y will be an integer. For example, all odd integer values of x generate integer values of y in the equation $5x + 2y = 3$. For more solutions, simply assign a different value to d (or m).

Incidentally, this problem has many solutions Diophantos would have found unacceptable because they are either: 1) trivial, that is (0, 3/2) and (5/3, 0) containing the non-number zero, or 2) solutions like (1, -1) such that either x or y must be negative and hence another non-number. His limited concept of numbers limited the solutions acceptable as valid. We, on the other hand, find negative integers, 0 and 1 perfectly acceptable in Diophantine solutions so we accept (1, -1), (3, -6), etc., although still we often prefer positive integer solutions. Note that accepting rational solutions, as did Diophantos, makes solutions like (1/5, 1) and (2/5, 1/2) acceptable to all.

Infinite Numbers of Solutions:

From the above solutions where d (or m) has an infinite range, it is obvious that all linear equations have an infinite number of solutions.

In addition, if any integer solution exists it can be used to generate an infinite set of integer solutions. That is, suppose (x_0, y_0) is an integer solution; then the ordered pairs $(x_0 + nb, y_0 - na)$ and $(x_0 - nb, y_0 + na)$ are obviously solutions as well, regardless of the value of n. Then, when n is any integer we have, as claimed, an infinite set of integer solutions.

Example Problem

A sheep farmer has a number of wool sheep and a few milk goats. She wants to sell some of each to pay a $5000 debt. The sheep are worth $80 each and the goats are worth $150. How many sheep and goats must she sell to get exactly $5000?

Answer: The relevant equation is . $80x+150y=5000$. Using the general solution $x = m$ and $y = \dfrac{500-8m}{15}$, it must be that

$$\frac{8m}{15} = n + \frac{1}{3} = \frac{5(3n+1)}{15}$$

where n is integer if y is to be integer. All one can do is raise the value of n incrementally to find values of $(3n + 1)$ that are multiples of 8. Thus we require odd values of n and the first solution occurs when n = 5 so that m = 10.

There are a number of solutions here for her to choose from: (10, 28), (25, 20), (40, 12) and (55, 4). Since she has few goats, the last of these will likely be the one she prefers.

Note that having a general solution is not the end of the challenge presented by a Diophantine problem.

Equations in Three Unknowns

The addition of a third unknown makes for a somewhat more challenging category of Diophantine problem. Suppose we are faced with an equation of the form

$$ax + by + cz = d \qquad (2)$$

Extending the method used for equations in two unknowns, assume $y = x + \gamma$ and $z = x + \delta$. With a little algebraic manipulation this gives the results

$$x = \frac{d - b\gamma - c\delta}{a + b + c}, \quad y = \frac{d - c\delta + a\gamma + c\gamma}{a + b + c}$$

$$\text{and } z = \frac{d - b\gamma + a\delta + b\delta}{a + b + c}$$

Assuming all the constants are integers, we then are guaranteed that, at least, rational solutions exist for any such problem.

Trivial Solutions:

For linear equations of the form of Eq. 2, where all the constants are known integers, we obviously have trivial, rational solutions of the form (d/a, 0, 0), (0, d/b, 0) and (0, 0, d/c). There are also three trivial rational solutions with only one variable equaling zero: (0, γ, (d – bγ)c), (-γ, 0, (d + a γ)/c) and (-δ, (d + aδ)/b, 0).

General Solutions:

We can simplify and make more specific the above general solutions by, for example, making γ = -1 and δ =

+1. Then these linear equations can always have solutions of the form

$$x = \frac{d+b-c}{a+b+c}, \quad y = \frac{d-2c-a}{a+b+c} \quad \text{and} \quad z = \frac{d+2b+a}{a+b+c}$$

Hence, every Diophantine linear equation of the above form has at least one general, non-trivial, rational solution. Neither this general solution nor the trivial solutions are certain to be integer solutions but is not difficult to generate integer solutions from them.

A word of warning is warranted, however. Generating an integer solution is likely to change d (in eq. 2) or c (in eq. 1). That is, the integer solution will not be a solution of the original equation! Consider a simple equation of the form of eq. 1: y = 3x − 2 with the solution (3/2, 5/2) when m = 3/2. Doubling the solution to get an integer solution gives (3, 5) which is not a solution of the original equation but of the related equation y = 3x − 4. That is, c must likewise double if we double the values in the rational solution. Of course, letting m = 2 generates the solution (2, 4) which is indeed an integer solution of the original equation but is not obviously derivable from (3/2, 5/2).

Infinite Numbers of Solutions:

As in the case of linear equations of two unknowns above, all linear equations in three unknowns which have at least one integer solution likewise have an infinite number of Diophantine (integer) solutions because if (x_0, y_0, z_0) is an integer solution, then the ordered pairs $(x_0 + nb, y_0 - na, z_0)$ and $(x_0 - nb, y_0 + na, z_0)$ are obviously integer solutions as well so long as n is integer. Thus, the range of the general integer solutions is also infinite.

Example Problem

A store keeper sells bedspreads for $120 each, sheet sets for $95 each and tablecloths for $45 each. At the end of a certain day her net income for the day was $3080. How many bedspreads, sheet sets and tablecloths did she sell?

Answer:
The relevant equation is . $120x + 95y + 45z = 3080$. Using the general solutions above, it must be that .

$$x = \frac{3080 - 95y - 45\delta}{200} \quad , \quad y = \frac{3080 - 45\delta + 105\gamma}{200}$$

and $$z = \frac{3080 - 95y + 215\delta}{200}$$

where n is integer if y is to be integer. Considering that $\frac{3080}{200} = 11\frac{11}{13}$ from the expression for x, it must be that

$$\frac{95y + 45\delta}{200} = j + \frac{11}{13} \quad \text{where } j \text{ is integer}$$

$$95y + 45\delta = 200j + 220$$

$$19y + 9\delta = 52j + 44 = 4(13j + 11)$$

Then it must be that $19\gamma + 9\gamma$ is a multiple of 4. This in turn forces γ and δ to be multiples of 4. Since γ = 4, δ = 4 does not yield an integer j, try γ = 4, δ = 8 for which we find j =

2. For the record, $\gamma = 8$, $\delta = 4$ generates $j = 36/13$, a non-integer values as does $\gamma = \delta = 8$. Going back into the general solutions we then find the solution (9, 13, 17). Since larger values soon drive x negative, we have evidently reached a good stopping place. We conclude the store keeper sold 9 bedspreads, 13 sets of sheets and 17 table clothes. Once again we see that having a general solution is only the beginning of getting a result.

Equations in More than Three Unknowns

The methods of the two previous sections can in principle be extended to four and more unknowns although the level of algebraic complexity naturally increases. It seems pointless, and would certainly be painful, to carry the discussion further.

The answer to Uncle Paul's problem is 36 geese. It is of course pointless to inquire how the lead goose became an English speaking mathematician.

Chapter 2. Quadratic Equations

Arguably the most famous and familiar Diophantine problems are quadratic equations in two unknowns. Solutions of the Pythagorean Theorem are merely the most famous of these but Pell's equation is another. Most of these do indeed have Diophantine solutions. For example, squaring the sides of eq. 1 yields a quadratic equation of the form

$$ax^2 + bxy + cy^2 = d \qquad \qquad 3)$$

without eliminating the status of the solution sets of the linear equation. Thus, it is certain that at least some quadratic equations have Diophantine solutions. Indeed, this argument easily extends to equations cubic, quartic and beyond. Eq. 3 has the trivial, but not certainly Diophantine, solutions $(\pm\sqrt{d/a}, 0)$ and $(0, \pm\sqrt{d/c})$.

Focusing for now solely on quadratics, we can also multiply two independent linear equations together to obtain the quadratic equation $(ax + by)(dx + ey) = c$ where c is now the product of two numbers. In this case, we are assured the quadratic has at least one solution because the solution sets of the two independent linear equations, being non-parallel straight lines, must intersect at a point in the xy plane. The values of x and y that make one equation generate an integer result cannot be expected to make the other equation do likewise so the single solution is the only one we can get, barring the trivial solutions $(\pm\sqrt{c/ad}, 0)$

and $(0, \pm\sqrt{c/be})$. None of these solutions are certain to be Diophantine.

For example, the equation $2x - y = 3$ has integer solutions $x = 3 + d$, $y = 3 + 2d$ where d is any integer and the equation $3x + 2y = 5$ has the integer solutions $x = 1 - 2n$, $y = 1 + 3n$ where n is any integer. It is easy to verify by substitution that these solutions generally do not work for their combination, $6x^2 + xy - 2y^2 = 15$. We can locate a solution at the intersection of the two lines by setting $3 + d = 1 - 2n$ and $3 + 2d = 1 + 3n$. This requires that $n = -2/7$ and $d = -10/7$ so that $x = 11/7$ and $y = 1/7$. Thus the quadratic has the rational solution $(11/7, 1/7)$, the intersection of the two original solution sets. Note that we have lost the integer solutions and are reduced to a merely rational solution. Of course, there are, in addition, the trivial but irrational and imaginary solutions $(\pm\sqrt{5/2}, 0)$ and $(0, \pm i\sqrt{15/2})$.

Probably the famous and important ancient problem of this sort is that of the Golden Mean. It is definitely not Diophantine. As is well known, it defines the proportions of the "most aesthetically pleasing rectangle" and is the basis on which much Greek architecture was framed. The rectangle has sides x and $x + y$ such that $x + y$ is to x as x is to y. From this we derive that $x^2 - xy - y^2 = 0$. By the quadratic formula, the ratio then has a single positive $(1+\sqrt{5})/2$ value. Obviously irrational, it is approximately 1.618034.

General Solutions:

Since the method above is neither obviously general nor Diophantine, are there solutions to these equations that we have missed? That is, is there a general method to find rational solutions of any particular example of eq. 3? The quadratic formula seems the obvious approach to any answer. First, since x and y will be numbers, there is no loss of generality in assuming $y = x + \beta$ where β is whatever rational number is required in any particular case. Then, by the quadratic formula

$$x = \frac{\beta(b + 2c) \pm \sqrt{\beta^2(b+2c)^2 - 4(a+b+c)(c\beta^2 - d)}}{2(a+b+c)}$$

If rational solutions exist, the discriminant must be the square of a rational number. For the above example ($6x^2 + xy - 2y^2 = 15$), we find a general solution

$$x = \frac{-3\beta \pm \sqrt{49\beta^2 + 300}}{10} \quad \text{and} \quad y = x + \beta$$

where the discriminant is always positive and rational so long as β is the root of a perfect square. If we let $\beta = \delta/7$ we need $\delta^2 + 300$ to be a square of a rational number or an integer and only $\delta = \pm 10$ satisfies this requirement. Thus, rational solutions exist for $\beta = \pm 10/7$ yielding (11/7, 1/7). as we already know, but also (-11/7, -1//7). This method provides a bit more insight than the first method, enabling us to see a second, related solution.

<u>Infinite Numbers of Solutions:</u>

We can simplify eq. 3 by dividing by the coefficient of x^2 and then rewrite it as $x^2 + bxy + cy^2 = d$ where the new coefficients are the old ones divided by a. Assuming we have two solutions, (x_1, y_1) and (x_n, y_n), the Brahmagupta identity then can be used to generate an infinite number of solutions by the recursion relations

$$x_{n+1} = (x_1 x_n - c y_1 y_n)/\sqrt{d}$$
$$y_{n+1} = (x_1 y_n + x_n y_1 + b y_1 y_n)/\sqrt{d}$$

If we now use these recursion relations on the above example, we find using either trivial solution with the other or with $(11/7, 1/7)$ or $(-11/7, -1//7)$ merely generates known solutions. However, using the solutions $(-11/7, -1//7)$ and $(11/7, 1/7)$ does generate two new, irrational, solutions which are approximately ± 1.56608, ± 0.28611.

For another example, the equation $x^2 + 3xy + 2y^2 = 28$ has the solutions $(1, 3)$ and $(-1, -3)$. Used in the above recursion relations we find another solution set: $(\pm 17/\sqrt{28}, \mp 33/\sqrt{28})$. We can obviously generate an infinite number of real solutions by this method but irrational solutions are generated copiously and rational solutions are unlikely except for equations where d is a perfect square and two rational solutions are available to begin the recursion process.

Example Problem

A farmer has a cultivated field 55 yards by 44 yards that netted him $12,100 last year. Anticipating a good market next year, he decides to double his net. Assuming the same return per square yard as this year, he must double his planted acreage. No mathematician, he realizes that means planting an area 110 by 44 yards or 55 by 88 yards. We are mathematicians so we can ask if there are other solutions.

Answer:

His yield per square yard was $12100/55 \times 44 = \$5$ and he wants to net $24,200. We can always add yardage to what is already there. In that case we have
$$(x+55)(y+44) = 24200/5 = 4840$$
But we can also write it xy = 4840 to simplify the problem. Let x = 55 and y = 88, or let x = 110 and y = 44 and we see the farmer's intuition was correct. Those are indeed possible solutions. It is worth noting in passing that this equation has no trivial solution.

Are there other integer solutions? Let y = x + d and we have $x^2 + xd = 4840$. To obtain integer solutions, the discriminant of the relevant quadratic equation must be a perfect square. That is, $d^2 + 4(4840) = D^2$ where d and D must both be integers. Unsurprisingly, solutions here are d = 33, D = 143 yielding x = 55, y = 88 and d = -66, D = 154 yielding x = 110, y = 44. Are there others? What about d = -33 or d = +66? They yield the solutions (-55, -88), (88, 55), (44, 110) and (-110, -44), none of which add any useful solution. It appears that there are no other Diophantine solutions.

However, the farmer can of course always accept an approximation and, for example, plant a 69.6 x 69.6 yards square field. Doubtless his greatest concern will be not the mathematics but the physical context, e.g., what land does he have that is contiguous to that already in use?

Simple Quadratics

We now lower our hopes and consider the simpler quadratic equations with d equal to a perfect square, say r^2. Possibly the simplest of these has the form

$$x^2 + d^2y^2 = r^2 \qquad \qquad 4)$$

where as yet we impose no restrictions on any of its parameters except that r and d must be integers.

Trivial Solutions:

Trivial solutions are then ($\pm r$, 0) and (0, $\pm r/d$). Because x and y appear only as squares, magnitudes are all that matter here. We can see from the trivial solutions that $r \geq |x| \geq 0$ and $r/d \geq |y| \geq 0$ for real solutions where as when $|x|$ increases $|y|$ decreases and *vice versa*.

General Solutions:

From the above quadratic formula solutions we can write solutions for eq. 4 in the form

$$x = A \pm \sqrt{A^2 + B}$$
$$y = x + \delta$$

$$\text{where} \quad A = \frac{\delta d^2}{d^2 + 1} \quad \text{and} \quad B = \frac{r^2 - \delta^2}{d^2 + 1}$$

For $0 < \delta < r$, A and B are positive which makes x and y real but not likely rational, let alone integer. Thus, we expect Diophantine solutions for eq. 4 to be uncommon.

Nonetheless, it is reassuring that real solutions are possible and we have established the required range of values of δ.

Seeking integer solutions, first note in eq. 4 that, as $|x|$ increases from 0 towards r, $|y|$ decreases from r/d toward 0. Hence the relevant range of $|x|$ values is 0 to r. Now, consider the general case where x = r − n where n is any integer. Substituting x into eq. 4 yields the result dy = $\pm\sqrt{n^2 + 2rn}$ so that y is integer if and only if $m^2d^2 = n^2 + 2rn$ where m is integer. To keep the x values in range we restrict n to the same range, 0 to r.

The problem then is to find integer solutions for m in the equation $m^2d^2 = n^2 + 2rn$ for the range $r \geq n \geq 0$. Solving for r we find r = ($m^2d^2 - n^2$)2n and we immediately see that the quantity $m^2d^2 - n^2$ must be even for integer solutions to occur. This is a quite severe restriction which means n and md must have synchronized parity, both odd or both even. In turn, this implies that m even requires n even. Warning! The reverse is not true; n even does not require m even but merely that at least one of the pair, m and d, must be even.

For any particular values of d and r, then, how many integer solutions occur? The answer is either 1 or 0 with 0 being the more common answer. Recurring to the equation $m^2d^2 = n^2 + 2rn$ and solving this as a quadratic in n we have that

$$n = \sqrt{r^2 + m^2d^2} - r.$$

The negative square root is ignored here to work only with positive n. For a positive integer solution for n the root must be an integer multiple of r, say $I_1 r$. This in turn implies that

$$nd = r\sqrt{I_1^2 - 1} = I_1 r \Rightarrow I_1^2 - I_1^2 = 1$$

and we have reached an impossibility. No two squares of integers differ by 1 except for the case 1 and 0 which entails the trivial solution n = 0 with either or both y = 0 and d = 0.

As a consequence of the above argument, we must abandon hope of finding integer solutions of eq. 4 for any arbitrary choice of d and r. However, the situation is not completely hopeless because, solutions do exist for selective choices of d and r. To see this, work with x = r − d rather than with x = r − n. Then $dy = \sqrt{2rd - d^2}$ which yields integer values of y so long as $2rd - d^2 = n^2 d^2$ where n is some integer. This helps us select r and d for they must be related as $2r = (n^2 + 1)d$. These conditions are easily fulfilled by many (r, d) pairs so long as either $(n^2 + 1)$ or d (or both) are even.

Finally, we have

$$y = n \text{ and}$$
$$x = (n^2 - 1)d/2 \text{ when } n^2 = (2r - d)/d \qquad 5)$$

Thus any odd n is possible (excluding n = 1 as trivial) and even n values may work so long as d is then also even. For example, the equation $x^2 + 4y^2 = 100$ makes n = ±3 and has the solution (8, ±3). As an example where n is even, consider $x^2 + 4y^2 = 25$ which makes n = ±2 and has the solution (3, ±2). Note that this procedure generates only one distinct positive integer solution for any particular values of d and r (and even then only for those equations fortunate enough to have r and d such that they produce a perfect square. As an example of failure of the method, consider the equation $x^2 + 9y^2 = 81$ with a non-integer n ($n^2 = 5$).

The above solution looks suspiciously like the famous Pythagorean triple (3, 4, 5). We will shortly consider the Pythagorean triples but, for the moment let us recheck the implications of our selection of d and r values. Back-substituting the x, y and r solutions into eq. 3 generates the equation

$$(n^2 - 1)^2 + (2n)^2 = (n^2 + 1)^2$$

which is transparently a particular case of the Pythagorean theorem (as well as an algebraic identity). Thus, the appearance of a Pythagorean triple solution should not be a surprise. Indeed, it implies that there are many other select choices of x = r − d leading to other Pythagorean triple solutions of eq. 4. For example, x = r −2d generates the triple (8, 6, 10) for n = 6, x = r − d/2 generates the triple (15, 8, 17) for n = 2 and x = r − d/3 generates the triple (8, 6, 10) again for n = 1, (35, 12, 37) for n = 2 and (80, 18, 82) for n = 3.

Infinite Numbers of Solutions:

Even with the above restrictions, it is obvious that the above general solution generates an infinite number of solutions of (an infinite number of) different examples of eq. 4 because it is well known that there are an infinite number of Pythagorean triples.

Attacking the question from a different direction, suppose now that we have at least two solutions of a particular equation designated (x_1, y_1) and (x_2, y_2). It is not difficult then to show by substitution that the set of recursion relations

$$x_{n+1} = x_1 x_n - \beta d^2 y_1 y_n$$
$$\text{and} \quad y_{n+1} = \beta y_1 x_n + x_1 y_n \qquad 6)$$

where

$$\beta = \pm \sqrt{\frac{x_1^2 - 1}{x_1^2 - r^2}} \qquad 7)$$

generates more solutions from the two known solutions. These are the previous recursion relations for eq. 3 recast into a somewhat different form. With no restrictions on the parameters, the solutions are also unrestricted; certainly we have no reason to expect them to be Diophantine. In fact, over the entire range $1 < x_1^2 < r^2$, β is imaginary. Hence, the choice of (x_1, y_1) requires caution.

Selecting the trivial solution $(r, 0)$ makes $\beta \to \infty$ but the other trivial solution, $(0, r/d)$, can be used with $(x_n, y_n) = (r, 0)$. A close look at eqs. 6 then reveals they must now generate $(x_{n+1}, y_{n+1}) = (0, r/d)$. In fact, using any combination of the trivial solutions merely generates another one of the trivial solutions. In essence, the trivial solutions are a closed group under the operations defined by eqs. 6 and 7. The trivial solutions, unfortunately, are easiest to obtain but are generally the least interesting.

To break out of the cycle, we need a non-trivial solution from the above general solutions. For example, consider the equation $x^2 + 4y^2 = 25$ with the trivial solutions (5, 0) and (0, 5/2). Here, n = 2 generates the obvious integer solution (3, 2). Take (0, 5/2) as (x_1, y_1) and let $(x_2, y_2) = (3, 2)$. Then $\beta = \pm 1/5$ and eqs. 6 generate a next rational solution set $(x_3, y_3) = (\pm 4, \pm 3/2)$. We have broken out of the trivial solutions. A second application of eqs. 6, still using $(x_1, y_1) = (0, 5/2)$, generates $(x_4, y_4) = (\pm 3, \pm 2)$ and we see that we have created a cyclic set of solutions which will not produce an infinite set of solutions even should we foolishly continue the process an infinite number of times. Restricting consideration to rational solutions where $r \geq |x| \geq 0$ and $r/d \geq |y| \geq 0$, we have only: (5, 0), (4, 1.5), (3, 2) and (0, 1.25). We can claim all these as Diophantine but that exhausts the possibilities. Each of these four, of course, defines a set of four when negative rational numbers are included.

Particular Cases of Equation 4

Requiring Diophantine solutions must impose restrictions on d and r and we can inquire as to the nature of such restrictions.

$d^2 = 0$

If $d^2 = 0$, Equation 1 reduces to $x^2 = r^2$ so that $x = \pm r$ and y may assume any value. Hence, an infinite set of Diophantine solutions exists for all integer values of r since y may be any integer. It is questionable, however, if there is any point in describing such a situation in terms of a variable, y, that does not actually appear in the defining equation. It seems more reasonable to simply say there are two Diophantine solutions for each value of r. In any event, the d = 0 case is too simple to be of much interest.

$d^2 < 0$

Negative values of d^2 are at least as interesting as positives; substituting x' = x/r and y' = y/r into eq. 4 generates a form of Pell's equation in x' and y' which is usually rewritten in x and y as

$$x^2 - fy^2 = 1 \qquad \text{Pell's Equation} \qquad 8)$$

where -f replaces d^2. This equation is factorable as $(x - y\sqrt{f})(x + y\sqrt{f}) = 1$. Thus, we can solve eq. 4 by first solving the allied form of Pell's equation. Note that integer solutions of Pell's equation must generate integer solutions of eq. 4 because r is integer.

If (x_1, y_1) and (x_2, y_2) are integer only and we now demand more solutions consisting only of integers, then, at the least, β must be real and integer. As eq. 4 indicates, these are demanding criteria. A rather simple case occurs when $x_1 = \pm 1$ so that β = 0. In this case, eqs. 6 reduce to

$$x_{n+1} = x_n x_n \quad \text{and} \quad y_{n+1} = x_n y_n \qquad 9)$$

and the recursion relations of equations 2 become

$$x_{n+1} = (\pm 1)^{n+1} \quad \text{and} \quad y_{n+1} = (\pm 1)^n y_n \qquad 10)$$

Then too, $y_1 = \pm\sqrt{(r^2-1)/d^2}$ which is only an integer if $\sqrt{r^2-1}$ is a perfect square integer multiple of d. For example, let $d^2 = 1$ and $r^2 = 5$. Then $y_1 = \pm 2$. We can then, for example, let $x_1 = -1$ and $y_1 = -2$ and $x_2 = 1$ and $y_2 = 2$. Eqs. 6 then generate $x_3 = -1$, $y_3 = -2$ and $x_4 = 1$, $y_4 = 2$. In fact, the solutions simply continue to alternate without generating any new solutions. The same result obtains if we choose $x_1 = -1$, $y_1 = 2$ and $x_2 = 1$, $y_2 = -2$. Alternatively, we might choose $x_1 = 1$, $y_1 = -2$ and $x_2 = -1$, $y_2 = 2$. Then $x_3 = -1$, $y_3 = 2$ and $x_4 = -1$, $y_4 = 2$ and so on *ad infinitum*. Likewise, the choice $x_1 = 1$, $y_1 = 2$ and $x_2 = -1$, $y_2 = -2$ then continually generates (-1, -2) *ad infinitum*. These results are merely what eqs. 9 imply. Choosing β = 0 is the controlling condition here because it forces us to use eqs. 6 which in turn forces $x_1 = \pm 1$. The choice $x_1 = -1$ leads to alternation between (x_1, y_1) and (x_2, y_2) as solutions whereas $x_1 = +1$ freezes the solutions on (x_2, y_2). Choosing different values of d and r merely change the value of y_1 without changing the overall behavior. For example, if d = ½ and $r = \sqrt{5}/2$, $y_1 = \pm 1$.

Solving Pell's equation

Around 628 AD, in one of the earliest works on Diophantine problems, the Indian mathematician and astronomer Brahmagupta invented a method of solving Pell's equation. Now called the *chakravala* method (the name implies it is cyclic, i. e., iterative) it depends greatly on Brahmagupta's identity in the form:

$$(a^2 + fb^2)(c^2 + fd^2) = (ac + fbd)^2 + f(ad - bc)^2$$
$$= (ac - fbd)^2 + f(ad + bc)^2 \qquad 11)$$

The identity shows that expressions of the form x + fy are closed under multiplication as they obviously also are under addition. The *chakravala* method can be outlined as follows.

1. Generalize Pell's equation to $x^2 - f y^2 = k$ (in Pell's equation k = 1).
2. Suppose there are many solution triples (x, y, k) and for this equation focus on two of them: (x_1, y_1, k_1) and (x_2, y_2, k_2).
3. Then by Brahmagupta's identity a new form of Pell's equation is

$$(x_1^2 - fy_1^2)(x_2^2 - fy_2^2) = k_1 k_2$$
$$= (x_1 x_2 + fy_1 y_2)^2 - f(x_1 y_2 + x_2 y_1)^2 \qquad 12)$$

which has the solution triple

$$(x_1 x_2 + fy_1 y_2, \ x_1 y_2 + x_2 y_1, \ k_1 k_2)$$

4. Now, start with the simple solution of Pell's equation where $k = m^2 - f$, (m, 1, m²-f), and

form with the solution triple (a, b, k) the new triple (am+fb, a + bm, k(m²-f)). It is important to keep in mf+2rind that m is, at present, an arbitrary and unassigned number. We find k by calculating it from assumed values of a and b.

5. Dividing the equation of the new triple by k^2 yields
$$\left|\frac{am+fb}{k}\right|^2 - n\left|\frac{a+bm}{k}\right|^2 = \frac{m^2-f}{k}$$

6. We are free to choose m so we choose it such that a) $\left|\frac{a+bm}{k}\right|$ is integer and b) m² - f is minimized, thus also minimizing $\left|\frac{m^2-f}{k}\right|$. We thereby generate a new triple.

7. Repeat steps 4-6 until a triple with k = 1 is found.

Example Problem

Solve $x^2 - 3y^2 = 1$

Start with the simple solution (m, 1, m2 – 3), where m is an unknown integer, and invent the triple (5, 2, k = 13). Requiring k = m2 – 3, we then have a new triple ((4a+3b)/13, (a+4b)/13, (m2 – 3)/12). To make (a+bm)/13 integer requires that (5+2m)/13 must be integer so make m

= 4 and we find that $(a + bm)/13 = 1$, $(am + fb)/13 = (20+6)/13 = 2$ and $(m2-f)/k = (16-3)/13 = 1$ and the new triple is the solution sought, $(2, 1, 1)$. This triple has both k = 1 and y = 1 so it is not only a solution of Pell's equation but is also the simple solution for this particular f value.

Once we have a solution in hand, can we generate more? Note that the trivial solution of Pell's equation, $(1, 0)$, is of no use in the recursion formulas of eqs. 6 and 7 since they can only return (x_n, y_n) for (x_{n+1}, y_{n+1}). However, these recursion relations do indeed generate more Diophantine solutions if we recognize that the simple solution, $(2, 1)$, is a stand-in for four solutions, namely $(\pm 2, \pm 1)$. Using pairs of these four solutions in eqs. 6 and 7 generates six new integer solutions: $(\pm 1, 0)$ and $(\pm 7, \pm 4)$. From these we can continue to generate new solutions. E. g., $(2, 1)$ and $(7, 4)$ generate both $(2, 1)$ again and the new solution $(26, 15)$. Then $(2, 1)$ and $(26, 15)$ generate both $(7, 4)$ and the new solution $(97, 56)$ and so on.

Another approach is to assume $x = f + r$. Substituting this into eq. 8 yields $y = \sqrt{f + 2r}$ must be integer so $y_n = n$ where n is any integer. This requires $n^2 = f + 2r$. We can summarize this: $y_n = n$, $x_n = y_n^2 + r - 2r^2$ and $2r^2 + f = n^2$. Since n can be, to this point, any integer, it seems there are obviously an infinite number of solutions. Actually n cannot assume just any integer value because of the requirement that $n^2 = 2r^2 + f$. The implication is that not every equation of the form $x^2 - fy^2 = r^2$ has solutions of this form. Thus, these solutions are not general to the equations

$x^2 - fy^2 = r^2$. When r = 1, these equations reduce to forms of Pell's equation with the solutions $y_n = n$, $x_n = y_n^2 - 1$ with the restriction that n must be integer and equal to $\sqrt{f+2r^2}$.

d² > 0

Positive integer values of d^2 begin with $d^2 = 1$, yielding undoubtedly the best known Diophantine problem: the Pythagorean Theorem for which r must also be integer. Eqs. 5 then become

$$y = n \text{ and}$$
$$x = (n^2 - 1)/2 \text{ when } n^2 = (2r - 1) \qquad 13)$$

For integer solutions n^2 must then be a perfect square, i. e., the square of an integer. Unfortunately, few integers qualify. It is easy to verify that those that qualify are only the odd integers: 5, 13, 17, 29, 37, 41, etc. Eqs. 13 then generate single positive solutions for each value of r each of which identifies four solutions when negative integers are taking into account.

Any integer multiple of the above solutions must also be a solution of the Pythagorean Theorem for r → rM where M is the multiplying integer. Thus the above list of odd integers does not exhaust the possible r values. For example, r = 5 leads to new equations where r = 10, 15, 20, etc.

Positive integer solutions of the Pythagorean formula are called Pythagorean triples (or triads), a, b, c, where a < b <

c. It has long been known that many triples can be generated directly from the relations: $b = p^2 - q^2$, $a = 2pq$ and $c = p^2 + q^2$ where p and q are positive integers, $p > q$. These are derived by use of Brahmagupta's identity mentioned above. Like the above method, this procedure tends to miss triples that are multiples of non-perfect square integers. For example, it will not generate (9, 12, 15) = 3(3, 4, 5).

The order $a < b$ may not always be retained with this formulation but c will be greater than a or b. When $p = q$, this formulation generates multiples of the triple (0, 1, 1), which might be considered Diophantine solutions. They are often overlooked because they describe no real triangles (and zero was initially not regarded as a number, let alone an integer).

This procedure generates Diophantine solutions for different integer values of r and will also generate an infinite series of solutions, albeit, non-Diophantine. If, for example, we choose (-3, -4) as (x_1, y_1) and (3, 4) as (x_2, y_2), then we get (x_3, y_3) and (x_4, y_4) as (-9 + 8i√2, -12 - 12i/√2) and (51 − 48i√2, 68 + 36i√2) respectively and so forth. Thus, we can generate an infinite series of complex solutions of the Pythagorean Theorem for any value of r by starting with the negative/positive pairs of any triple. Although there are sometimes more than one triple for a particular value of r, no value of r attaches to a series of solutions so we designate no recursion relations for Diophantine solutions of the Pythagorean formula.

Example Problem - Simultaneous Equations Yielding a Quadratic

Find an x and y such that their sum and the sum of their squares are the given numbers m and n. As Diophantos presents it, m = 20 and n = 208.

Thus, $x + y = m$ and $x^2 + y^2 = n$. Eliminating y we find $2x^2 - 2mx\ m^2 - n = 0$. By the quadratic formula then

$$x = \frac{2m \pm \sqrt{4m^2 - 4(2)(m^2 - n)}}{4} = \frac{m \pm \sqrt{2n - m^2}}{2}$$

This result obviously puts restraints on n and m because the discriminant must be a positive square number. The choice above, m = 20 and n = 208 is judicious, making x = 12 or 8. Diophantos noted both.

Pythagorean triples

Positive integer solutions of the Pythagorean formula $c^2 = a^2 + b^2$ are called Pythagorean triples, a, b, c where usually a < b < c. Triples that are not integer multiples of other triples are known as *primitive triples*. As noted previously the expressions $a = 2pq$, $b = p^2 - q^2$ and $c = p^2 + q^2$ generate many of the triples. Of course, p and q are positive integers. The values of p and q are restricted by three conditions: p > q, p and q are relative primes and one of the two must be odd and the other even. The order a < b may not always be retained with this formulation but c will be greater than a or b. The trivial solution 0, n, n fits in here with q = 0, but it is generally ignored because it describes no real triangle. I do not include it for consideration. Pythagoras, of course, did not recognize zero as a number (nor did he call 1 a number either; it was, for him, the basis of numbers and not a number in itself). It is important to note that, from the restrictions listed, the relations will not usually generate perfect squares.

Note that the conditions on p and q above mean that a must be a multiple of 4. We can use that fact to create a generator of Pythagorean triples other than (p,q). Require that a = 4n where n is any positive integer, b = d + f and c = a +d where d = $f^2/(2(4n-f))$. This move creates the new generators (n, f) replacing (p, q). n = d = m generates m times the triple (3,4,5). All other values of n generate more than one, not always unique, triple. This set is not as efficient a generator as (p, q) but it has the important benefit of strongly emphasizing the fact that at least one element of a triple must be a multiple of 4. That fact leads to the following theorem.

<u>Lemma 2.1:</u> *For every integer multiple of 4 there is a corresponding Pythagorean triple where the other two elements are odd numbers.*

Proof: From the (p, q) relations, a = 2pq. Then, the requirement that p and q be an odd-even pair insures that a contains 2 factors of 2 and is therefore a multiple of 4. For q = 1, then p is even, say = 2n where n is any integer. Then a = 4n, the set of all multiples of 4. The odd-even requirement also implies that, of the pair p^2 and q^2, one must be odd and one even. From the relations defining b and c then, we see they must both be odd because $p^2 \pm q^2$ must always be odd.

This lemma does not apply to every Pythagorean triple. For example, it obviously applies to the triple (4, 3, 5) but does not apply to the triple 2·(4, 3, 5) = (8, 6, 10). The a = 8 triple to which it applies is (8, 15, 17).

Note that the q = 1, p = 2n condition sets up a set of triples for every multiple of 4 where c − b is always 2. Examples of this are (4, 3, 5), (8, 15, 17) and (12, 35, 37). There are, of course, other combinations of 2pq that generate multiples of 4 so the above set does not exhaust the list of triples with a as a multiple of 4.

<u>Lemma 2.2:</u> *No more than one of the three elements of a Pythagorean triple can be the square of an integer.*

Proof: The familiar triple (3, 4, 5) shows that one element of a triple can be the square of an integer. To show that no more than one can be a square, we can try squaring a by requiring $p = 2q$ but then $b = 3q^2$ and $c = 5q^2$ and neither is a square. Similarly, if we make $b = d^2$ then $c = 2p^2 - d^2 = (\sqrt{2}p + d)(\sqrt{2}p - d)$ and $a = 2p(p^2 - d^2)^{1/2} = 2p(p + d)^{1/2}(p - d)^{1/2}$ so neither a nor c can be squares. Lastly, if we make $c = d^2$ then $b = 2p^2 - d^2 = (\sqrt{2}p + d)(\sqrt{2}p - d)$ and $a = 2p(d^2 - p^2)^{1/2} = 2p(p + d)^{1/2}(p - d)^{1/2}$. As before, neither a nor b can be squares.

Comment 1: *"No squares at all"* is the rule for the vast majority of Pythagorean triples. As p >> q both b and c, of course, approach p^2 as a limit but never truly become squares.

Comment 2: A Pythagorean triple is a set of integers and a set of rational numbers and any set of rational numbers that solves the Pythagorean Theorem, say a/b, c/d and e/f where all parameters are integers, on multiplication by bdf generates a Pythagorean triple adf, cbf, ebd. Hence, integer solutions are equivalent to rational solutions and vice versa.

$d^2 > 1$

If we now consider $d > 1$ and integer, we are back to the topic of the previous section. In a move we did not make in that section, we can define $w = dy$ so that eq. 4 becomes, again, the Pythagorean formula in the variables x and w which has the solutions examined above for x and y. The triples also apply here but dy must now account for the y value of the triple. Hence, for example, $x^2 + 16y^2 = 25$ has the Diophantine solution $x = 3$ and $y = 1$ from the triple (3, 4, 5) but the equation $x^2 + 15y^2 = 25$ has no Diophantine solutions. In general, we can let $x = p - q$ where $p > q$ and $y = 2\sqrt{pq}/d$. Then $r = p + q$ where p and q can be made relative primes integers. So x and n are certainly integers but only if $q = 1$ and p is a squared integer. In the equation above then, $y = 4/15$ and a non-trivial integer solution is not possible.

Pythagorean Kin

Closely related to the Pythagorean equation and, now breaking the pattern we have been following of only considering quadratics with the RHS a perfect square, is the equation

$$x^2 + y^2 = n \qquad \qquad 14)$$

Of course it has Pythagorean triples as solutions when n = m² but it has integer solutions for other values of n as well; for example, $13 = 3^2 + 2^2$ and $10 = 3^2 + 1^2$. In fact, it has integer solutions so long as (x, y) comprise an integer pair. However, there are no integer solutions for certain values of n.

Solving it generally, we assume y = x + a. This sets up a quadratic equation in x which easily produces the general solution that x = (b – a)/2 and y = (b + a)/2 where $2n = a^2 + b^2$. There are no restrictions on a and b except that they must be integers and have the same parity. For the examples above then, for the first n = 13, a = -1 and b = 5 and for the second n = 10, a = -2 and b = 4. Note that negative integers have a natural role to play here and a, b and n are of the same parity.

Unfortunately, the solution in (a, b) is as complex as the original problem of finding (x, y). Its sole advantage over (x, y) is the requirement of having the same parity. The choice of a and b then generates n, x, and y. The procedure is the reverse of finding a solution; it *begins* with the solution and finds the corresponding equation.

On the other hand, the procedure gives us insight into why there is no solution for certain values of n. For example, there is no Diophantine solution for n = 14 because 14 is never the sum of just two squares, it requires at least three squares, $14 = 3^2 + 2^2 + 1^2$. This observation suggests a new problem; to wit:

$$x^2 + y^2 + z^2 = n \qquad \qquad 15)$$

Certainly $x^2 = y^2 = z^2 = n/3$ is a solution here although if n = 0 it is a very trivial one. However, I will pass over this problem in favor of a related one of more historical interest. Diophantos did indeed attack similar problems and I will take on one of his own examples in Chapter 5.

It has been known since antiquity that any positive integer, n, equals the sum of no more than four square integers. That is

$$w^2 + x^2 + y^2 + z^2 = n \qquad \qquad 16)$$

where some of the unknowns may be zero. This is a challenging problem that is usually solved by trial and error. A complicating factor is that values of n usually have multiple solutions. The trial and error method can be considerably facilitated by forethought. None of the unknowns can exceed $\sqrt{n}$ and the largest unknown cannot be less than about $\sqrt{n/2}$. Thus, we have a search range that limits the trials for the largest unknown. Having estimated the largest unknown, square it and subtract from n. The square root of the remainder then provides an estimate of the next largest unknown, etc. Using some simple programming on a modern computer can swiftly generate all possible solutions for any particular value of n.

I will close the chapter with a few remarks about the rather odd equation

$$x^2 + y^2 = z^3 \qquad 17)$$

which has the trivial solutions (1, 0, 1) and (0, 1, 1). Using Brahmagupta's identity (BI)

$$(p^2 + q^2)^3 = (p^3 + pq^2)^2 + (p^2q + q^3)^2$$
$$= (p^3 - 3pq^2)^2 + (3p^2q - q^3)^2 \qquad 18)$$

we obtain general solutions of two forms: $z = p^2 + q^2$ and then either $x = p^3 - 3pq^2$, $y = 3p^2q - q^3$ or $x = p^3 + pq^2$ and $y = p^2q + q^3$ where p and q can be any integers including zero. Making p or q be zero yields the trivial solutions above. The choice $p = q = 1$ generates the solutions (2, 2, 2) and (-2, -2, 2). If $p = 1$ and $q = 2$, $z = 5$ and (x, y) = (-11, -2) or (5, 10) and so forth. With no limits on the integers p and q, we see immediately that the equation has an infinite number of solutions.

The related equations

$$(n-1)x^2 + y^2 = z^3 \text{ and } x^2 + (n-1)y^2 = z^3 \quad 19)$$

both have the solutions (±n, ±n, n) where n is any integer greater than 1.

Chapter 3. Useful Lemmas for Higher Orders

We have so far made little use of Brahmagupta's identity (BI) except for recursion relations and for solving the Pythagorean Theorem and eq. 18. Working with higher order equations requires higher powers of manipulation. Although it has perhaps not yet been readily apparent, the identity is a powerful tool. Irrational and complex numbers also arise quite naturally in algebraic solutions but Diophantine analysis rejects them. I think it worthwhile, then, to develop some tools that will be of considerable assistance if one aims to extend Diophantine analysis beyond this beginning. I will present them in a series of lemmas.

Lemmas on Primes and Square Numbers

<u>Lemma 3.1:</u> Let α, β, a and b be integers and let $N = \alpha^2 + \beta^2$ (the sum of two integral squares). Also let $r = N/p$ where r is integer and $p = a^2 + b^2$ is a prime integer, then r is also the sum of two integral square numbers.

Proof: First, multiply r, top and bottom, by $a^2 + b^2$

$$r = \frac{\alpha^2 + \beta^2}{a^2 + b^2} = \frac{(\alpha^2 + \beta^2)(a^2 + b^2)}{(a^2 + b^2)^2}$$

Now, applying the BI

$$r = \frac{(\alpha a \pm \beta b)^2 + (\alpha b \mp \beta a)^2}{(a^2 + b^2)^2}$$

$$= \left|\frac{\alpha a \pm \beta b}{a^2 + b^2}\right|^2 + \left|\frac{\alpha b \mp \beta a}{a^2 + b^2}\right|^2$$

Hence, r is the sum of two squares. Are the squares also integers? Consider the first square. Because of the ±, the numerator represents two squared integers. We hope at least one of these contains a factor of $a^2 + b^2$. If that is so, the product of the two numerators will also be divisible by $a^2 + b^2$. Forming the product we find

$$(a^2\alpha^2 - b^2\beta^2) = a^2(\alpha^2 + \beta^2) - \beta^2(a^2 + b^2)$$
$$= (ra^2 - \beta^2)(a^2 + b^2)$$

The product indeed contains $a^2 + b^2$ as a factor so that at least one of the two possible squares is integer. The second set of squares is susceptible of the same treatment with the same result.

$$(b^2\alpha^2 - a^2\beta^2) = b^2(\alpha^2 + \beta^2) - \beta^2(a^2 + b^2)$$
$$= (rb^2 - \beta^2)(a^2 + b^2)$$

Thus, at least one pair of the two pairs of squares will be integer. But r is an integer so the other of the pair must also be integer, QED

<u>Lemma 3.2:</u> *Let α, β, a and b be integers such that $p = a^2 + \gamma b^2$ is a prime number and $N = \alpha^2 + \gamma\beta^2$, then if the ratio $r = N/p$ is integer, then it is also of the form $r = c^2 + \gamma d^2$ where c and d are integers.*

Proof: All that is required here is an obvious modification of the above proof of Lemma 3.1 by making $b^2 \to \gamma b^2$ and $\beta^2 \to \gamma\beta^2$.

<u>Lemma 3.3:</u> *No prime number can equal two different pairs of squared integers.*

Proof: The proof is essentially by contradiction. We assume we have a prime number that is simultaneously equal to two different sets of square integers: $p = a^2 + b^2 = c^2 + d^2$ where a and b are not equal to c and d. We can then write p^2 as

$$p^2 = (a^2 + b^2)(c^2 + d^2) = a^2c^2 + a^2d^2 + b^2c^2 + b^2d^2$$

A little algebraic manipulation shows the cross terms below all cancel so this is equivalent to

$$p^2 = (ac + bd)^2 + (ad - bc)^2$$
$$= (ac - bd)^2 + (ad + bc)^2$$

We can also manipulate the two last terms above

$$(ad - bc)(ad + bc) = a^2d^2 - b^2c^2$$
$$= a^2(c^2 + d^2) - c^2(a^2 + b^2) = p(a^2 - c^2)$$

Comparing the RHS and LHS of the line above indicates that p must be a factor in either (ad - bc) or (ad + bc).
All primes are odd except for 2 for which we know $2 = 1^2 + 1^2$ is the only solution. Then one member of the pairs a, b and c, d must be odd and the other even. Let a and c be even and b and d be odd. Then, (ac - bd) and (ac + bd) are both odd and, hence, neither one is zero. From the middle line above then, both (ad - bc) and (ad + bc) are less than p. But one of them has p as a factor so it can only be p times zero. Therefore $p(a^2 - c^2) = 0$ forces $a^2 = c^2$ which then requires that $b^2 = d^2$. The assumption of different sets is then false and only one set is possible, as asserted by the lemma.

Comment: The lemma does not imply that all primes equal the sum of only two squares. For example, $31 = 5^2 + 2^2 + 1^2 + 1^2 = 3^2 + 3^2 + 3^2 + 2^2$.

Lemmas on Irrational Numbers

<u>Lemma 3.4:</u> *The product and the sum of any two irrational numbers are either irrational or prime integers.*

Proof by Contradiction: Let a and b be irrational numbers and by assumption we require their product a·b = l/m where l and m are relative prime integers. Rewriting the equation as m·(a·b) = l we find that the product must be an integer for l to be integer. That means that, contrary to assumption, m is an integer factor of l. The product must be irrational.

Similarly, consider the sum. As before, by assumption a + b = L/M where L and M are relative prime integers. Let a = ln(c) and b = ln(d) so the sum then becomes ln(c·d) = L/M or M·ln(c·d) = L. Since L and M are integers so must ln(c·d) be an integer. That makes M an integer factor of the integer L which contradicts the assumption that L and M are relative prime integers. Therefore the assumption that a + b is rational is necessarily false. The sum is irrational.

The above conclusions can be escaped iff M or m equals 1. M = 1 implies L is prime and m = 1 implies l is prime.

Lemma 3.4 has great power as the following corollaries demonstrate.

Corollary 1: *The expressions $\pi \pm e$ and $\pi \cdot e$ are all irrational numbers.*

The conclusion follows because $\pi \pm e \approx 3.14159 \pm 2.71828 \approx 5.85987$ or 0.42331, neither of which can be a prime number, and $\pi \cdot e \approx 3.14159 \times 2.71828 \approx 8.35972$ which also cannot be prime.

Corollary 2: *The expressions $\pi \pm \phi$ and $\pi \cdot \phi$ are all irrational numbers.*

The conclusion follows because $\pi \pm \phi \approx 3.14159 \pm 1.618034 \approx 4.759624$ or 1.523556, neither of which can be a prime number, and $\pi \cdot \phi \approx 3.14159 \times 1.618034 \approx 5.08320$ which also is not a prime.

Corollary 3: *The expressions ϕ ± e and ϕ·e are all irrational numbers.*

The conclusion follows because ϕ ± e ≈ 1.618034 ± 2.71828 ≈ 4.33631 or -1.100245, neither of which is a prime number, and ϕ·e ≈ 1.618034 x 2.71828 ≈ 4.39827 which also cannot be prime.

The rationality of linear combinations of irrationals, for example aπ + be, is currently uncertain so the following theorem is of real interest.

<u>Lemma 3.5:</u> *If a and b are different integers and i and j are the magnitudes of irrational numbers, linear combinations of the form ai + bj are always irrational.*

Proof by Contradiction: Assume it is true that ai + bj = m/n where m and n are relative prime integers so that m/n is rational. Let ai = ln(c) and bj = ln(d). Then ai + bj = ln(c·d) and the assumption becomes ln(c·d) = m/n. Then n·ln(c·d) = m and ln(c·d) must be an integer. But then n and m are not relative primes and the assumption is false. Therefore, the coefficients are irrational rather than rational.

<u>Lemma 3.6:</u> *For n any positive integer not the square of a positive integer, the square root of n is irrational.*

Proof by Contradiction: Assume $\sqrt{n}$ is rational. Then $\sqrt{n}$ = l/m where l and m are positive integers and relative primes. Squaring gives $nm^2 = l^2$. If n is a perfect square of the integer c, then l = mc is the integer product of two positive integer and there is no contradiction so long as m=1. If not, however, l^2, being a square, must contain 2 factors of the integer n. That is, l = nd where d is another positive integer. The equation then reduces to $m^2 = nd^2$. By the same argument then we see that it must be true that m = ne where e is another positive integer. At this point we have a contradiction of the initial assumption that l and m are relative primes.

Corollary of Lemma 3.6: *If n is a positive rational number not the square of a positive rational number, the square root of n is irrational.*

<u>Lemma 3.7:</u> *For n and k > 1 positive integers with n not the kth root of a positive integer, the kth root of n is irrational.*

Proof by Contradiction: Assume $\sqrt[k]{n}$ is rational. Then $\sqrt[k]{n}$ = l/m where l and m are positive integers and relative primes. Raising this to the kth power gives $nm^k = l^k$. If n = c^k, then l = mc is the integer product of two positive integer and there is no contradiction so long as m = 1. If not, however, l^k must contain k factors of the integer n. That is, l = nd where d is another positive integer. The equation then reduces to $m^k = n^{k-1}d^k$. By the same argument then we see that it must be true that m = ne where e is another positive integer. At this point we have a contradiction of the initial assumption that l and m are relative primes.

<u>Lemma 3.8:</u> *Any integer root of an irrational number is irrational.*

Proof by Contradiction: Assume the nth root of an irrational number, i, is rational. Then $\sqrt[n]{i} = 1/m$ where l and m are integers. Raising the equation to the n power gives $i = l^n/m^n$. Both l^n and m^n are integer when l and m are integer so we conclude that an irrational number is rational. The contradiction shows the assumption false and the theorem *true*.

Corollary of Lemmas 3.7 and 3.8: *If a^2 is irrational, then a and $a^{2/n}$ (n integer) are the square and n = n integer roots of an irrational and are also irrational.*

<u>Lemma 3.9:</u> *Any integer root (n^{th} root) of a non-integer rational number, p/q, is irrational so long as p and q are not n^{th} powers of integers.*

Proof: Assume the nth root of a rational number, p/q, is rational. Then $\sqrt[n]{p/q} = 1/m$ where p and q, l and m are relative prime pairs of integers and neither m nor q can equal 1. Raising the equation to the nth power gives $p/q = l^n/m^n$. If $p = l^n$ and $q = m^n$, the equation is an identity and there is no contradiction. A rational but trivial solution also occurs for $q = l^{-n}$ and $p = m^{-n}$. There are no other rational solutions. Therefore, any other solution is, perforce, irrational.

Chapter 4. Cubic Equations in Two or More Unknowns

Simple Cubics

Surely the best known cubic equation in Diophantine analysis relates to Fermat's Last Theorem:

$$x^3 + y^3 = z^3 \qquad 20)$$

It is important not for its solutions but because it has been proved that it has no integer solutions. See, for example, Carmichael's version (*Diophantine Analysis*, p. 67).[*]

A somewhat more complex equation with a wide range of Diophantine solutions is

$$x^3 + y^3 + z^3 = w^3 \qquad 21)$$

which has an infinite number of trivial solutions of the form (n, 0, 0, n) and their permutations but has also the interesting infinite set of solutions (3n, 4n, 5n, 6n). In fact, any solution of the equation can be multiplied by n to generate an infinite number of solutions. Other solutions

---

[*]Robert D. Carmichael, Diophantine Analysis, John Wiley & Sons, Inc., New York, 1915.

are: (11n, 15n, 27n, 29n), (18n, 19n, 21n, 28n), (27n, 30n, 37n, 46n) and (29n, 34n, 44n, 53n).

Meanwhile the similar looking equation

$$x^3 + y^3 + z^3 = w^2 \qquad \qquad 22)$$

has the trivial solution (1, 0, 0, 1) and its permutations and the (almost) trivial solution (3, 3, 3, 9). Other solutions are: (1, 2, 3, 6), (6, 14, 16, 84), (8, 13, 15, 78), (9, 14, 16, 87), (15, 37, 38, 330), (18, 24, 30, 216), (20, 25, 30, 225), (20, 25, 39, 288), (22, 26, 28, 224), (23, 24, 25, 204), (34, 40, 41, 415), (50, 56, 57, 697), (50, 64, 73, 881), (54, 59, 61, 768), (58, 66, 81, 1007). Actually, since by symmetry x, y, and z can be permuted, each of these solutions is one of six related solutions.

These solutions all generate an infinite number of related solutions by a simple extension. For example, (3, 3, 3, 9) generates the solution set (3m2, 3m2, 3m2, 9m3) where m is any integer.

Theorem 4.1: *There are no positive integer solutions of* $z^3 = x^3 + y^3$.

Proof: We may regard this problem as that of finding the roots of $z^3 - x^3 - y^3 = 0$ where we seek only positive integer solutions. As above, we will order the unknowns $z > y > x$ and we may rewrite this condition in the form $y = x + \kappa$ and $z = x + \lambda$ where $\lambda > \kappa$ and neither is necessarily integer. This move immediately generates the cubic equation

$$-x^3 + 3x^2(\lambda - \kappa) + 3x(\lambda^2 - \kappa^2) + \lambda^3 - \kappa^3 = 0 \qquad 23)$$

The solutions of a cubic are well known to depend on the value of a discriminant of the form

$$D = \frac{b^2}{2} + \frac{a^3}{27} \quad \text{where here } a = 6\lambda(\kappa - \lambda)$$

$$\text{and } b = 3\lambda(\lambda - \kappa)(\kappa - 2\lambda) \qquad 24)$$

Some algebra quickly finds that

$$D = \frac{9}{2}\lambda^2(\kappa - \lambda)^2(\kappa - 2\lambda)^2 + 8\lambda^3(\kappa - \lambda)^3 \qquad 25)$$

$$D = \lambda^2 \left| \frac{9}{2}\kappa^4 + 10\lambda^3 + \frac{59}{2}\kappa^2\lambda^2 - 19\kappa\lambda(\kappa^2 - \lambda^2) \right| \qquad 26)$$

Keeping in mind that $\lambda > \kappa$, we see that D is positive. This means that Eq.1 has only one real solution of the form $x = A + B$ where

$$A = \left| -\frac{b}{2} + \sqrt{D} \right|$$

$$\text{and} \quad B = \left| -\frac{b}{2} - \sqrt{D} \right| \tag{27}$$

Since b is negative, -b/2 is positive as is D but the magnitude of D is greater than that of b/2. Thus $B < 0$ while $A > 0$ and $|A| > |B|$. Therefore, x is positive but is it integer? By Lemma 3.6 above, the square root of any positive integer is irrational unless the integer is a perfect square. Thus, D must be a perfect square of a half integer if x is to be integer. But it is never a perfect square because the quartic expression in Eq.26 is not a perfect quartic and therefore not a perfect square. By Lemma 3.7 and its corollaries A and B are then both irrational and the sum of two irrationals must be irrational. This means x is always irrational and is never integer. QED.

Theorem 4.2 (Half of Fermat's Last Theorem): *There are no positive integer solutions of $z^{2n} = x^{2n} + y^{2n}$ where n is an integer > 1.*

Proof by Contradiction: If we rewrite $z^{2n} = x^{2n} + y^{2n}$ as $(z^n)^2 = (x^n)^2 + (y^n)^2$ we have the Pythagorean Theorem and assuming x, y, z form an integer solution of the equation, the values x^n, y^n, z^n must also be integer. But (x^n, y^n, z^n) must be a Pythagorean triple and cannot have more than one element as the square of an integer; the other two must be non-squared integers. Hence, the set $x^{n/2}$, $y^{n/2}$, $z^{n/2}$, at best is 1 integer and 2 irrationals by Lemma 2.2. By Lemma 3.7, then, taking the 2/n root of the set elements, the elements x, y, z might at best be one integer and 2 irrationals. Assuming positive integer solutions leads to a contradiction and the theorem is true.

Theorem 4.3: *If x and y must be integers, the equation $x^y = y^x$ has only one non-trivial solution.*

Proof: There is obviously an infinite number of trivial solutions of the form x = y. If x ≠ y, then assume x = cy where c must be an integer. The equation becomes $c^y y^y = y^{cy}$ which implies that $c^y = (y^y)^{c-1}$ or $c = y^{c-1}$. From this we conclude that $y = {}^{c-1}\sqrt{c}$. The zeroth root with c = 1 has no meaning so, starting with c = 2, we find y = 2 and x = 4. This is the only non-trivial integer solution because the function ${}^{c-1}\sqrt{c}$ with c integer never reaches another integer but drops from 2 to approach 1.0000 as c approaches infinity. To make the claim of approach to 1 analytic, consider the limit over the range 2 to ∞, of

$$\ln \left| \lim x^{-1} \right| = \lim \left| \frac{\ln x}{x-1} \right|$$

The logarithm and the limit may be inverted in order as shown because the function is continuous over the relevant range. Applying L'Hôpital's Rule the RHS reduces to

$$\lim \left| \frac{1/x}{1} \right| = \lim(1/x) = 0$$

Inserting this result into the first expression gives as asserted.

$$\lim \left| x^{-1} \right| = 1$$

Theorem 4.4: *There are no positive integer solutions of $z^{4n} = x^{4n} + y^{4n}$ where n is an integer > 0.*

Proof by Contradiction : Assume the equation has positive integer solutions x, y, z. We rewrite the equation as $(z^{2n})^2 = (x^{2n})^2 + (y^{2n})^2$. Then x^{2n}, y^{2n} and z^{2n} are positive integers and must form a Pythagorean triple. By construction, these terms are all squares but, by Theorem 4.2, this is impossible. Therefore, the set x^{2n}, y^{2n}, z^{2n} is not a set of squared positive integers; no more than one of the three can be a squared integer. Taking square roots, only one of the three numbers x_n, y_n, z_n might be an integer and, by Lemma 3.6, the other two are irrational numbers. Now taking the nth roots and applying Lemmas 3.6, 3.7 and 3.8, it follows that only one of the three values x, y, z might be rational, the other two must be irrational. Unfortunately, we have assumed they are all integer. By contradiction then, positive integer solutions of the equation do not exist and the theorem is true.

Example Problem
Diophantos tackled just a single cubic problem in the *Arithmatika*. His version, translated literally, is oddly stated. I reframe it as follows: find a right triangle such that 1) its perimeter is a cubic number and 2) its areal value added to the value of one of the perpendicular sides is a square number. 3) as a right triangle, the sides fit the

Pythagorean theorem $a^2 + b^2 = c^2$ and (a. b, c) can only be a Pythagorean triple.

Solution:

Let the sides be a and b and the hypotenuse be c. Then we have

    1) $a + b + c = A^3$
    2) $a + ab/2 = B^2$.
    3) $a^2 + b^2 = c^2$.

As so far stated, this is a "three equations in five unknowns" type system, essentially an impossible problem. Diophantos added another condition to aid us. Stripped of its verbosity, he required that $A^3 - B^2 = 2$. Seemingly he has simply added a fourth equation to the initial three. That is not so because he knew very well the equation has a single solution in the integers. Without admitting it he has effectively said, "Let $A = 3$ and $B = 5$." To extend the illusion of deep insight, he made $A = m - 1$ and $B = m + 1$ and solved for m.

$$(m - 1)^3 - (m + 1)^2 = 2 \quad \rightarrow \quad m^3 - 4m^2 + m - 4 = 0$$

The obvious solution is that $m = 4$. His only cubic is hardly challenging. Then, of course, $A = 3$ and $B = 5$ as we already recognized. The apparently nasty problem turns out to be something of a set up for Diophantos to show off.

Finding (a, b, c) remains as a problem of three equations in three unknowns but Diophantos was ready with another simplification. He made the area of the triangle = x = ab/2. Hence ab = 2x. This inspired the move <u>b = 2</u>, a = x. Condition 2 then becomes 2x = 25 and 1 becomes c = 25 − x. That makes condition 3 → $x^2 + 4 = (25 − x)^2 = 625 − 50x + x^2$ or <u>a = x = 621/50</u>. Then the solution triple is (100/50, 621/50, 629/50).

Chapter 5. Other Equations in Two or More Unknowns

Diophantine problems that do not fit the equation categories used so far are easy enough to discover but more difficult to systematize. I will not try to do so but content myself with presenting exemplars.

Equations with Inverse Powers

A category of Diophantine problems that do not obviously belong anywhere is those occasional problems with negative powers of unknowns. As a general category it has no limit but in actual fact Diophantos attempted few of these.

Example Problem

Consider then the following problem from *Arithmatika*:

A wealthy man had a number of sons but only two daughters. The bulk of his estate had to go to his sons but his daughters needed dowries. He decided then to devote only an nth of his estate to the dowries with one daughter receiving an xth of it and the other daughter a yth. Algebraically, the problem is now

$$\frac{1}{x} + \frac{1}{y} = \frac{1}{n}$$

For n = 4, Diophantos was aware of the solutions: x = y = 8, x = 5, y = 20 and, x = 6, y = 12. How is such an equation generally handled?

Answer

First, rationalize it by multiplying by the lowest common denominator of all the fractions. Here that yields nx + ny = xy, a quadratic equation. Next, reduce to a single unknown plus unknown constants by letting y = x + k. The result, 2nx + nk = x^2 + kx, can then rewritten into the familiar quadratic form x^2 + (k - 2n)x − nk = 0. Applying the quadratic formula we find that

$$x = \frac{2n - k \pm \sqrt{4n^2 - 4nk + k^2 + 4nk}}{2} = \frac{2n - k \pm \sqrt{4n^2 + k^2}}{2}$$

For integer solutions, $4n^2 + k^2$ must be a perfect square. If k = 0 the requirement is automatically fulfilled, leading to the obvious solution that x = y = 2n. Note that in general, k = 2n is an upper limit on solutions beyond which x is driven into negative values.

More generally, it must be the case that $m^2 = 4n^2 + k^2$. The only integer solutions are then Pythagorean triples. If, for example, the father considers designating a twentieth of his estate to dowries, he can then give each daughter a fortieth of his estate. Alternatively, he might give one daughter a 39th and the other a 42nd (n = 20 and k = 3). The only other possibility is x = 30, y = 60 where k = 30.

Systems of Linear Equations

Systems of linear equations are now the domain of linear algebra but they can generate problems that have Diophantine solutions. Diophantine or not, modern matrix methods are the go to procedure for solving them. However, a few Diophantine examples will do no harm.

For simple situations such as two equations in two unknowns, the standard trick of let y = x + k works readily. Consider the following two equations with integer coefficients: ax + by = e and cx + dy = f. They become (a + b)x = e - bk and (c + dk)x = f − dk. On eliminating y between them we find that

$$x = \frac{(e - bk)}{(a+b)} = \frac{(f - dk)}{(c+d)}$$

From which we deduce that

$$k = \frac{f(a+b) - e(c+d)}{d(a+b) - b(c+d)}$$

By restricting the numbers a through f to integers we have also then restricted k to the rational numbers. Under those conditions, the problem is Diophantine in the sense that the solutions are rational. If we desire solutions where x and y are both integers, the numerator must then be an integer multiple of the denominator, a much more restrictive, but certainly not an impossible, condition.

Now recall that true Diophantine problems are under-determined but we have so far been considering a fully determined system. Thus, a problem where we know only one of the two equations constitutes a true Diophantine problem.

Example Problem
A farmer sells some sheep at $30 a head and some goats at $40 a head receiving a total of $260. How many sheep and how many goats did he sell?

Answer: We can write the transaction as $30x + 40y = 260$.
The answer $x = 2$ and $y = 5$ demands almost immediate consideration but suppose we are obtuse and do not see it. Let $y = x + k$ and the equation becomes $70x = 260 - 40k$
Or $7x = 26 - 4k$. Making the right hand side a multiple of 7 allows x to be an integer. Now 21 and 7 do not fit in but14 does fit when $k = 3$. So then the solution is $x = 2$ sheep and $y = 5$ goats, as we saw at the outset.

Since we did all the above work, can we now generate a second equation in x and y? In addition to the values of x and y we have $a = 30$, $b = 40$, $e = 260$ and $k = 3$. Can we find c, d, and f? Well, that involves us in another true Diophantine problem. The above equations reduce to just a single equation, $2c + 5d = f$. Unfortunately the problem is greatly under determined. Possible solutions are: $c = 1$, $d = 1$ and $f = 7$ with $x + y = 7$; $c = 2$, $d = 1$ and $f = 9$ with $2x + y = 9$ and so on. Obviously, there is no limit to the number of equations in x and y that suit this condition.

Archimedes' Cattle Problem

Archimedes of Syracuse (c. 287 – c. 212 BC) on Sicily was the greatest mathematical physicist of antiquity. Justly proud of his invention of a way of expressing large numbers in terms of powers of 10,000 (a myriad), he celebrated the achievement by showing how to calculate the number of sand grains needed to fill the universe. Realizing that a heliocentric universe would be larger than a geocentric one, in his Sand Reckoner, he naturally had to give an answer for both systems.

The following problem is widely attributed to him. Given his interest in problems involving enormously large numbers, the attribution seems likely to be correct.

Statement of the Problem

Compute the number of the cattle of the sun which once grazed upon the plains of Sicily, divided according to color into four herds, one milk-white, one black, one dappled and one yellow. The number of bulls is greater than the number of cows, and the relations between them are as follows:

White bulls = (1/2 + 1/3) black bulls + yellow bulls,
Black bulls = (1/4 + 1/5) dappled bulls + yellow bulls,
Dappled bulls = (1/6 + 1/7) white bulls + yellow bulls,
White cows = (1/3 + 1/4) black herd,
Black cows = (1/4 + 1/5) dappled herd,
Dappled cows = (1/5 + 1/6) yellow herd,
Yellow cows = (1/6 + 1/7) white herd.

Find the number of each kind of bulls and cows.

To make the problem even more formidable, we can add the following two conditions:

White bulls + black bulls = a square number,

Dappled bulls + yellow bulls = a triangular number.
(The triangle numbers are: 1, 3, 6, 10, 15, 21, 28, 36, 45, 55 etc.).

Answer

The first part of the problem has no apparent connection to Pell's equation. It is a system of seven linear equations in eight unknowns. Assigning obvious symbols to the different quantities, we can write the following seven equations.

W = (5/6)B + Y
B = (9/20)D + Y
D = (13/42)W + Y
w = (7/12)(B + b)
b = (9/20)(D + d)
d = (11/30)(Y + y)
y = (13/42)(W + w)

The system is, of course, indeterminate and it has infinitely many solutions. To solve it required much tedious effort. We proceed either by eliminating variables by substitution or solving by matrix methods.
Starting with the first three equations we can fairly quickly deduce that

$$Y = \frac{2673}{6078} H$$

$$D = \frac{2497701}{2110078} H$$

$$B = \frac{1009206}{2110078} H$$

The four remaining equations can be reduced to the single equation

$$w = W + \frac{529200}{27027}B + \frac{238140}{27027}D + \frac{1044351}{(12)(27027)}Y$$

in which we can now eliminate B, D and Y for the result that

$$4(27027)(6678)w = W \begin{vmatrix} 4(27027)(6678) + 529200(4800) \\ +23814(4740) + 2072(348117) \end{vmatrix}$$

As it transpires, the least set of positive integers satisfying the seven equations is:
 B = 7,460,514
 W = 10,366,482
 D = 7,358,060
 Y = 4,149,387
 b = 4,893,246
 w = 7,206,360
 d = 3,515,820
 y = 5,439,213

a total of 50,389,082 cattle. The other solutions are integral multiples of this set, obviously an infinite number of possibilities. All in all, a herd this size would greatly stress the "plains of Sicily."

The solution of the second part of the problem was not found until 1880; it is that the above solution set of the first part of the problem should be multiplied by

$$w = \frac{(u^{2658} - u^{-2658})^2}{(4657)(79072)}$$

where

$$u = 300426607914281713305\sqrt{609}$$
$$+ 84129507677858393258\sqrt{7766}$$

and *j* is any positive integer.

Equivalently, squaring *w* yields

$$w^2 = s + t\sqrt{(609)(7766)}$$

where (s, t) are solutions of the Pell equation

$$s^2 - (609)(7766)t^2 = 1$$

The smallest solution of this equation (for j = 1) has far too many digits (≈ 10^5) to write out here but it can be given as approximately 7.76 x 10^{206544}. That is far more than the number of sand grains needed to fill a heliocentric cosmos. In fact, the current universe has a volume of about 10^{79} cubic meters and could not possibly hold this herd of the "cattle of the sun."

Systems of Double Equations

Diophantos consistently used the term "double equations" to refer to a pair of equations of the general form $F(x) = a^2$ and $G(x) = b^2$.where the functions are often first order but may also be second order. The following two examples are taken from his *Arithmatika*.

Example 1:

Divide a given number, n, into three parts, x, y and z, such that 1) $x + y + z = n$, 2) $xy + z = u^2$ and 3) $xy - z = v^2$.

Solution:
This is a set of three equations in three unknowns and can be systematically solved as such. After expending effort, Diophantos was able to recognize the latter two equations as a pair of double equations for which he already had worked out solution procedures. Modern methods are more systematic so I will proceed with them rather than try to explain the older method. Eliminating xy from 2 and 3, we readily find that $\underline{z = u^2 - xy}$, $z = (u^2 - v^2)/2$ and $\underline{xy = (u^2 + v^2)/2}$. We now write 1) as $x + y + u^2 - xy = n$ to find that $\underline{x + y = n + (v^2 - u^2)/2}$ and can eliminate y between the last two underlined equations to find the quadratic in x,

$$2x^2 - (v^2 - u^2 + 2n)x + u^2 + v^2 = 0$$

which can be solved easily once we know u, v, and n. For example, Diophantos worked with $n = 6$, finding that $u = 48/9$, $v = 1/9$, $x = y = 5/3$ and $z = 8/3$. You will note the quadratic equation then becomes factorable as $(3x - 5)^2 = 0$ so that $x = 5/3$. The remaining parameters can then be found by the above equations and are indeed the values Diophantos bequeathed us.

Example 2:

Find three ordered numbers, x, y and z, such that x + y, x + z and y + z are all squares of rational numbers and furthermore that z − y = 3(y − x). Organizing these conditions in modern form we have:

1) $z > y > x$
2) $z - y = 3(y - x)$
3) $x + y = a^2, y + z = b^2, x + z = c^2$

Solution:

Diophantos began by combining conditions 3) and 1): x + y = a^2 and y > x implies that y > a^2/2. Then let y = a^2/2 + w so that x = a^2/2 − w. We can then solve condition 2) for z to find z = a^2/2 + 7w. It also follows that w = (y − x)/2. From condition 3) we now find that

$$b^2 = 8w + a^2$$
$$c^2 = 6w + a^2$$

from which we see that $b^2 > c^2 > a^2$. The problem is thereby reduced to finding x, y, z and c^2 and b^2 from the parameters a^2 and w.

Before following Diophantos,, note that eliminating w results in the equation $a^2 + 3b^2 = 4c^2$ plus the corresponding equation z + 3x = 4y. With the quadratic we revert to equation 4 of Chapter 2 above with d^2 = 3 and r = 2c. That 3 is not a perfect square rules out Pythagorean triples here but, since Diophantos posed the problem, presumably it has a Diophantine solution.

The method Diophantos used was a substitution we have already used in finding Pythagorean triples. Begin with the fact that $8w + a^2$ and $6w + a^2$ are squares and make the substitutions:

$$8w + a^2 = (p + q)^2$$
$$6w + a^2 = (p - q)^2$$

Subtracting the second line from the first yields $2w = 4pq$ which can be factored as $2p = w/2$ and $2q = 4^*$. Hence, $8w + a^2 = (w/4 + 2)^2 = (w^2 + 16w + 64)/16$. On multiplying by 16 we obtain $\underline{w^2 = 112w + 16\,a^2 - 64}$. Applying the quadratic equation and varying a between 1 and 1020, there are only 7 solutions with integer values of the parameters a and w. They are listed in the table below.

a	w	x	y	z	a^2	b^2	c^2
2	112	-110	114	786	4	900	676
8	120	-88	152	872	64	1024	784
13	132	-47.5	216.5	1008.5	169	1225	961
22	160	82	402	1362	484	1764	1444
47	252	852.5	1356.5	2868.5	2209	4225	3721
383	1593	71751.5	74937.5	84495.5	146689	159433	156247
44	910	58	1878	7338	1936	9216	7396
2	1.88016	0.11983	3.88016	15.1611	4	19.0413	15.2810

Of course, the first three solutions could not satisfy Diophantos because of the negative values of x. He pointed out that we are dealing with three squares: $8w + a^2$, $6w + a^2$ and a^2 whose differences are obviously related as $b^2 - c^2 = 8w$. This a simple Pell's equation with the obvious solution that $b = 3\sqrt{w}$, $c = \sqrt{w}$. Is there an a^2 such that w is

---
********* Other factorings produce the same result.

a perfect square? Scanning the w column above we see there are no such w values available.

What Diophantos did next was involved and ingenious. I will elaborate it rather than explain it. The first solution in the top row above (with a = 2) is the only one of the 6 he seemed aware of. But the implication that x = -110 is most unsatisfactory. He therefore retained a = 2 (now see the bottom row of the table) but accepted rational fractions for all the other results. I have transformed this into the row above that by multiplying by 22^2 (except that the new a is 2 x 22). He thus arrived at an entirely positive set of integers for x, y and z as well as a full set of three perfect squares for a^2, b^2 and c^2. A bravura performance!

All the solutions above may be divided by perfect squares to generate additional rational solutions. Hence, the solution set of the problem is infinite. But we have not yet exhausted the possible integer solutions. The simplest solution of the problem is x = y = z = 2 for which a = b = c = 2 also. Dividing x, y and z by any perfect square then also generates a solution in rational numbers (fractions of course). Thus, we have an additional infinity of solutions here too. Diophantos took no note of this. Indeed, it could never have occurred to him because it violates the first condition. Admittedly, condition 1 is implicit, rather than explicit, in his statement of the problem.

Various Sets of Simultaneous Equations
Example 1.
Sets of equations of different order are not unusual. An example from the *Arithmatika* is actually quite simply

solved in modern algebraic form. Stated verbally it is: find two numbers whose sum is A and the sum of their squares is B. Hence, we write:

$$a) \quad x + y = A$$
$$b) \quad x^2 + y^2 = B$$

Eliminating y from a) and substituting it into b), we have

$$x^2 + (A - x)^2 = 2x^2 - 2Ax + A^2 = B$$

Solving this quadratic with the quadratic formula yields

$$x = \frac{2A \pm \sqrt{4A^2 - 4(2)(A^2 - B)}}{4} = \frac{1}{2}(A \pm \sqrt{2B - A^2})$$

Then also y = A − x and the problem is solved. Obviously, considering the discriminant, 2B must be greater than A to avoid complex or imaginary x and y values. For rational solutions, the discriminant must be a perfect square so A and B must be selected carefully. Knowing Pythagorean triples comes in handy here. Consider (6, 8, 10). Make A = 6 and 2B = 100 so that the discriminant is 8. Then x = 7 or -1 and y = -1 or 7.

Example 2.

Also from *Arithmatika* we have the following: find two numbers such that the square of either added to the other is a square. Hence, we write:

1) $x^2 + y = a^2$ and   2) $x + y^2 = b^2$

Eliminating x or y here is a poor choice because it results in a quartic equation. But since we know $(x + 1)^2 = x^2 + 2x + 1$, let $y = 2x + 1$ so that $x + 1 = \pm a$, satisfying condition 1. Then $x = \pm a - 1$, $y = 2a - 1$ and condition 2 becomes $(2a)^2 = b^2 \pm 3a$. Should we choose to work with $(y + 1)^2$ instead, as one would expect, we get $(2b)^2 = a^2 \pm 3b$.

The obvious solution for either equation is $a = b = 1$ from which we obtain the trivial solutions (1, 0) and (0, 1). It is readily apparent that (a, 0) and (0, b) are also trivial, but more general, solutions.

However, the forms of the equations make one think of Pythagorean triples. Using +3 in the first equation and letting $a = m/k$ and $b = n/k$, the equation now becomes $(2m)^2 = n^2 + 3mk$ and we can force the last term to be a square that makes a Pythagorean triple an infinite number of possible solutions. For example, try to force the triple (6, 8 10) making m = 5 and n = 6. Then k = 64/15, a = 75/64, b = 90/64, x = 11/64 and y = 86/64.

Diophantos with his usual ingenuity finds the solution x = 3/13 and y = 19/13 which certainly works but does not fit any Pythagorean triple!

Simultaneous Equation of Higher Order

Diophantos proposed some problems in equations of order higher than cubic. A particularly clever one is to find two numbers, x and y, such that: 1) $x^3 + y = a^3$ and 2) $x + y^2 = b^2$.

Solution:

A straight-forward attempt at eliminating variables immediate creates sixth powers. Diophantos had a clever move available, however.

He let $b = x^3 + a^3$ so that $b^2 = x^6 + 2a^3x^3 + a^6$. Eq.. 1 becomes $y = a^3 - x^3$ and 2) becomes

$$x + a^6 - 2a^3x^3 + x^6 = b^2 = x^6 + 2a^3x^3 + a^6$$

which simplifies to $x = 4a^3x^3$. Then we have $x^2 = 1/4a^3$. Then x^2 is a perfect square if $\underline{a = 4}$ for which $\underline{x = 1/16}$. From 1), $\underline{y} = 64 - 1/16^3 = \underline{262143/4096}$.

www.ingramcontent.com/pod-product-compliance
Lightning Source LLC
Chambersburg PA
CBHW070428180526
45158CB00017B/920